Amazon Tap

Ultimate User Guide to Mastering Your Amazon Tap

Andrew Mckinnon

© Copyright 2016 - All rights reserved.

In no way is it legal to reproduce, duplicate, or transmit any part of this document in either electronic means or in printed format. Recording of this publication is strictly prohibited, and any storage of this document is not allowed unless with written permission from the publisher. All rights reserved.

The information provided herein is stated to be truthful and consistent, in that any liability, in terms of inattention or otherwise, by any usage or abuse of any policies, processes, or directions contained within is the solitary and utter responsibility of the recipient reader. Under no circumstances will any legal responsibility or blame be held against the publisher for any reparation, damages, or monetary loss due to the information herein, either directly or indirectly.

Respective authors own all copyrights not held by the publisher.

Legal Notice:

This book is copyright protected. This is only for personal use. You cannot amend, distribute, sell, use, quote or paraphrase any part or the content within this book without the consent of the author or copyright owner. Legal action will be pursued if this is breached.

Disclaimer Notice:

Please note the information contained within this document is for educational and entertainment purposes only. Every attempt has been made to provide accurate, up to date and reliable complete information. No warranties of any kind are expressed or implied. Readers acknowledge that the author is not engaging in the rendering of legal, financial, medical or professional advice.

By reading this document, the reader agrees that under no circumstances are we responsible for any losses, direct or indirect, which are incurred as a result of the use of information contained within this document, including, but not limited to, —errors, omissions, or inaccuracies.

Table of Contents

Introduction 7
Chapter 1 Setting Up Your *Amazon Tap* 15
Chapter 2 How It Works 25
Chapter 3 Features 35
Chapter 4 Specifications 51
Chapter 5 Alexa 55
Chapter 6 Security Concerns While Using Alexa 67
Chapter 7 General Precautions to Remember 73
Chapter 8 Troubleshooting 81
Conclusion 87

Andrew Mckinnon

Introduction

I want to thank you and congratulate you for downloading the book, *Amazon Tap*.

This book provides all the information that you need to use *Amazon Tap* to the best of its abilities. From proven steps and strategies that can help you maximize the use of *Amazon Tap* to details on how you can lengthen its use, you can be sure that the ideas you need are all available in this guide.

Music is a vital part of most people's lives. There are some who cannot get through the day without listening to a song. Some choose to hear songs all the time while working because they believe it makes them more productive. Many studies prove this theory is right. Music can give people a lot of benefits, so it is not surprising that products related to music are always evolving and are extremely popular items.

When it comes to bringing your music with you wherever you go, it's true that our phones can fulfill this purpose, but their speakers may prove insufficient in many instances, even those on high-end devices. On the other hand, it is possible to hook up various external speakers to your phone via a cord, of course, but these are often impractical, as we all know the troubles associated with cables. Bluetooth technology has been bolstering the advance of wireless connectivity for some time now, and Bluetooth speakers are certainly among the most practical innovations in this field.

If you are acquainted with devices similar to *Amazon Tap* or the Echo, you will have noticed that Bluetooth speakers have been heading in a new direction over the past few years. And if you haven't been paying attention to the market, you will certainly see what this means by the time you are finished reading this book. Namely, the times of these speakers merely serving the purpose of playing music is going away fast. Instead, such devices are now fitted with an artificial intelligence of sorts, the likes of which are Siri, Alexa, Google Now, Cortana, and others. They play the role of a voice assistant and are really the next step in developing hands-free options for personal devices available to all.

The addition of these voice assistants to wireless speakers, in particular, has seen them begin to evolve into much more than they would otherwise be. Devices such as the *Amazon Tap,* Echo, or Dot can now play the role of a control unit in your home. Through their voice control systems, you can run your home's security or inner workings, such as doors, lights, thermostats, or any other piece of smart technology you may have installed – all simply through spoken words. And, of course, while doing all this, you can listen to your favorite tracks and genres on the speaker. Add to all this the portability of a wireless speaker, and you've come out with something along the lines of *Amazon Tap.*

Searching for the right wireless speaker used to be an easy task because of the limited selection available but many innovations in speakers have changed the standards for listening quality. The Amazon Echo is a huge hit among speaker fans, so consumers are also receiving the release of Amazon Tap positively.

The Amazon Echo is considered too expensive by a lot of

consumers, and the fact that it cannot be moved from one place to another is deemed to be a hindrance by others. *Amazon Tap*'s battery-powered feature, allow it to be portable made many speaker fans excited about its release. This feature will probably explain why you decided to purchase an *Amazon Tap*. If your goal is to listen to music whenever you want while at the same time doing a variety of other things, the *Amazon Tap* is a perfect choice for you. You may have become accustomed to just using portable speakers for listening to good music without realizing that maybe this particular speaker can do a lot more.

This portable Bluetooth and Wi-Fi-enabled speaker can provide great sound that will allow you to listen to your favorite songs. This feature is considered different from all other brands of speakers because, instead of having just one side available, it can provide 360-degree sound because the speaker extends all around. It received its name mainly because you can just tap on it, provided that it is connected to any Wi-Fi network, and you can ask for music, search online for possible information, and hear the latest news. *Amazon Tap* is one speaker that is different from all the others that you have tried before. The *Amazon*

Tap is similar to the original Echo, but with an emphasis on being a portable and lightweight social experience. Yes, the *Amazon Tap* does not need to be plugged in constantly, so you can take it on the go.

The *Amazon Tap* is available with various color skins, making it a personalized style accessory. With a height of just 6.2 inches and a weight of only 470 grams, the *Amazon Tap* is portable and easy to move around. Whether you need to place it somewhere inside your home or you want to bring it with you, it will not be hard to do. You can also charge it to use it for a few hours. With a full charge, the *Amazon Tap* can play up to 9 hours of music or last up to three weeks in standby mode. If you need to use it for a longer period, or if the battery is running low while you are using it, you can use the cradle charger, which charges *Amazon Tap* while you use it. Using the *Amazon Tap* will ensure that you will stay entertained for a longer span of time.

Wondering how the *Amazon Tap* is so different from all the other available speakers? The difference is mainly because of the Alexa Voice Service, the application that controls *Amazon Tap*. Although Alexa Voice Service is not unique to this device, as it also powers other Bluetooth speakers created by Amazon, such as Echo and Echo Dot, the *Amazon Tap* will still give you what the other two speakers can offer. Alexa as a virtual assistant is also being highly modified so that it can carry out more commands and requests.

In this book, you will get all necessary information that you need to make *Amazon Tap* work well for you. You will be able to gather details to maximize the use of this portable gadget. This book is also full of basic knowledge that you should not forget, because it will allow you to understand what exactly to do. This guidebook is designed to give you the good, the bad, and never-before-written details about *Amazon Tap*. After reading this handbook, you will realize that you now have all the information required to master the *Amazon Tap* and the Alexa Voice Service. It is very likely that, after reading this book, you will understand that purchasing the *Amazon Tap* is one of the greatest decisions that you have made. You will also be more confident about using it because of the information that you have gained.

Of course, you are encouraged to research further if you feel that something was omitted despite the depth of detail in this book. One reason to look up those things that occur to you after reading through the book is the swift advancement and growth of software such as Alexa. One of the most important things you will learn here is that new ways in which you can make your voice assistant useful are popping up all the time.

Thank you for downloading this book! Hopefully, you will enjoy it as much as you will enjoy using your *Amazon Tap*!

Andrew Mckinnon

Chapter 1
Setting Up Your *Amazon Tap*

Setting up the *Amazon Tap* is very similar to setting up similar products, such as the Echo and Echo Dot. Most people purchase *Amazon Tap* because it is similar to the Echo but more portable. Setting up *Amazon Tap* can be easy to do, even if you have no previous knowledge about setting up products.

Take some time to read the information illustrated below and, even if you have no experience with using a Bluetooth speaker, you will feel like a pro. Follow our simple steps, and you will be able to set this up in no time at all!

What's in the Box?

The *Amazon Tap* box contains these components:

- Amazon Tap
- Charging Cradle
- Micro USB Cable and 9W Power Adapter

- Quick Start Guide

Unpacking

Please make sure that you unpack this from all the protective gear. Carefully remove the protective layer to minimize the risk of damage to the speakers. Take extra caution when using sharp objects for cutting the protective gear.

Keep in mind that, unlike Amazon Echo, your *Amazon Tap's* exterior is covered in a more vulnerable, fabric material, which is far more susceptible to being cut by a sharp object like a knife or a box cutter. However, this doesn't mean that the Tap is too easy to damage in general; what's more, the exterior design of the Tap is not prone to scratches because of its cover, which means its appearance will remain unaltered for longer unless it's dropped, hit, or cut.

Battery

The battery of the *Amazon Tap* can last for up to 9 hours for playback or 3 weeks for standby mode. The battery is rechargeable, which means that you do not need to replace it every time the power. The battery life will depend on various factors that cannot be controlled by the speaker.

You can also use *Amazon Tap* while it is charging so that you can keep listening to great music and it can continue doing your commands without interrupting or slowing down. Please charge your *Amazon Tap* for a minimum of 2 hours before you begin to use it.

Although it's apparent, it is worth noting that while the cradle is a very convenient charging system that a lot of users like. It will make your Tap stationary, making it more like the Echo, which is always stationed in one spot and plugged into the wall. This is also the one considerable drawback to this device, as you will still have to press the microphone button to issue commands to Alexa, even when the Tap is sitting in its charging dock. This is a disadvantage because the Echo's Alexa is listening for your commands all the time. Regardless, the fair longevity of the battery itself and countless other features that we will cover will definitely outweigh this weakness.

Setting Up *Amazon Tap*

Now that you have checked all the things that came with your *Amazon Tap*, it is time to learn how to set it up so that you can use it appropriately. Setup can be a breeze as long as you follow these simple steps:

1. Download Alexa Operating System

Alexa is the brain behind *Amazon Tap*. It is a built-in cloud-based voice service that enables you to give voice-powered commands to your device, provided it has a valid Internet connection. We will cover Alexa in more depth later in this book. You can download the Alexa application for various web browsers. It does not matter whether you are a Fire user, iOS user, or Android user, it can be compatible, provided that you have downloaded the required versions for your smartphone or gadget. Please be guided accordingly:

- Fire OS 2.0 or higher for Fire users is downloadable from the Amazon AppStore.
- Android 4.0 or higher for Android users is downloadable from Google Play.
- IOS 7.0 or higher for iOS users is downloadable from the Apple App Store.

If you are going to download Alexa to your computer, please make sure that your computer is connected to the Internet. Also, make sure that you have an updated web browser, or you will not be able to download the Alexa application correctly.

If you have an older version of software than those mentioned above, you may have to update it. The update will not take a lot of time. Please remember that some devices may not support the Alexa application.

2. Starting *Amazon Tap*

You need to turn on the device before you can continue setting it up. It is recommended that you place it on the charging cradle and plug it in so that *Amazon Tap* will be powered up. You will know that the *Amazon Tap* is on when the light indicators light up (first, the *Amazon Tap* will turn blue and then the light color will change to orange). Alexa will give its greeting and the setup can continue.

3. Connecting to the Wi-Fi Network

You need to make sure that the *Amazon Tap* will connect to the Wi-Fi network. Otherwise, setup will not continue. The process may automatically start, and you just have to follow the given instructions but, in case the installation does not start automatically, search for the Wi-Fi button located on the gadget and press it for about 5 seconds. To use the Alexa app on your phone or gadget, please go to settings and choose the option, "Set up a new device." If it does not work the first time you try it, you may have to do it again. Others who have problems setting up the *Amazon Tap* to connect with Alexa might have to restore the *Amazon Tap* to its original settings first.

You will do almost all of the connecting through the Alexa app on your smartphone by simply following instructions. Once the Tap is up and running, your phone will recognize it, and you can pair the two devices through your Wi-Fi settings, where you will find that the Tap has been detected – the app's instructions will lead you to this. After the Tap has been connected to your phone, it is time to connect it to your actual Wi-Fi network at home or wherever you may be so that it can gain access to the internet. This is essentially the whole process, and your *Amazon Tap* should now be ready to put in work.

4. Start Using *Amazon Tap*

1. Once you are sure that you have connected your *Amazon Tap* to your device, you can start using your speakers and talk to the Alexa Voice Service. Unlike other Amazon speakers, you first need to tap the microphone button on the *Amazon Tap* before you talk. While this is a slight inconvenience, it is simply

because the *Amazon Tap* doesn't stay plugged in like the other Amazon speakers, and having Alexa turn off when you don't need it conserves battery life.

5. **Using the *Amazon Tap* as a Bluetooth Speaker**

To stream music from your mobile device to the *Amazon Tap*, you will first need to pair them. Before you can do this, you have to set your mobile device to Bluetooth pairing mode and ensure that it is within the range of your *Amazon Tap*. It is then a simple two-step process to connect and a one-step process to disconnect your *Amazon Tap*. You will need to pair your *Amazon Tap* before you are able to successfully stream your music.

1. Begin by pressing the Wi-Fi/Bluetooth button on your *Amazon Tap*; Alexa will let you know when your *Amazon Tap* is ready to pair.
2. Open the Bluetooth settings on your mobile device, make sure that Bluetooth is enabled, and select the *Amazon Tap*. Alexa will tell you when the connection is successful. Once it is successful, you will be ready to stream your audio from the source of your choice.
3. To disconnect your *Amazon Tap* from your mobile device, press the power button on the *Amazon Tap*.

Some Tips to Remember

- Although you can link your *Amazon Tap* to your phone, Alexa is not able to receive your phone calls, text messages, and other notifications from your mobile device. She also cannot send audio from your *Amazon Tap* to Bluetooth speakers or headphones.
- Once your mobile device has been connected to your *Amazon Tap* once, you can simply turn on the Bluetooth on your mobile device, press the microphone button on your *Amazon Tap* and say "Connect." If you have connected to multiple mobile devices, it will connect to the most recently paired device.
- When it comes to giving commands, it will be best if you keep them short and straightforward. If *Amazon Tap* does not recognize the command the first time, rephrase it. After some time, you will realize what words you need to say to get what you are looking for. You can also teach Alexa new skills that were not in her initial programming. We will cover this in more depth in Chapter 5: Alexa.
- If you want to preserve battery life, place the item in sleep mode. You can do this by pressing the power button without holding it. When your *Amazon Tap* goes

into sleep mode, the power button will no longer be lit.
- The device will warn you when the battery is low by speaking so you can put *Amazon Tap* on charge instantly and you do not have to check its battery status from time to time.
- You can also see if the battery is low based on the color flashed by the indicator lights.

Where to Place

The best thing about *Amazon Tap* is that you can place it anywhere. If you are at home, you can put *Amazon Tap* right wherever you think it is appropriate. You can also carry *Amazon Tap* with you wherever you travel. Imagine having *Amazon Tap* with you while you are on the beach with your family members and friends or if you have an outdoor party at home.

You can set it up where it will be safe and audible at the same time. The right location will allow you to get the best out of *Amazon Tap*. Due to its cylindrical shape, you can even put *Amazon Tap* in the bottle rack of a bicycle. If you need to bring *Amazon Tap* with you, you may choose to carry it inside your bag. It will not take much space. Similarly, you may also consider purchasing a sling, which will allow you to clip it on the items you are carrying.

Remember that your *Amazon Tap* disperses sound 360 degrees around itself; this is also very convenient when it comes to finding a place to put it, as it provides you with more options. You can put it next to something that blocks one or two sides of the Tap, and the sound will still be crisp and clear. Plus, you never have to worry about delicate positioning or which way the device is rotated; you can put it down as you would a bottle.

One thing to keep in mind when you are choosing the spot to place your *Amazon Tap* is that you want it to be somewhere that is safe. Don't place your *Amazon Tap* on a ledge where it could be prone to getting bumped and falling. Also, don't place it near water or anywhere it can get wet or suffer other damage. Since the *Amazon Tap* is so portable, you are sure to find the perfect place for it with minimal trial and error.

Chapter 2
How It Works

Now that you have set up *Amazon Tap*, and have decided how you are going to use it, the next question is: How it will work? *Amazon Tap* is a device that you can just tap and ask. It will come up with an answer and will give you what you are searching for.

Please keep in mind few simple directions for more precise use of *Amazon Tap*:

- Make sure that *Amazon Tap* is turned on. You can do so by observing the lights that appear on the top right corner of the *Amazon Tap*.
- Press the microphone button every time you need to state a command. It is important to remember that, unlike some other Alexa-enabled devices, when you are using the *Amazon Tap*, Alexa cannot hear you if you do not press the microphone button.
- You may find that there will be some modifications as

Alexa becomes updated. Updated versions of Alexa will make *Amazon Tap* much smarter and better to use each time. You can also manually take Alexa's sophistication even further by downloading Alexa Skills to expand her assortment of smarts. It's also a great way to tailor your *Amazon Tap* to your specific, personal needs. We will go into more detail on this and touch upon some examples later in this book.

Hardware Basics

You need to understand some basic controls of *Amazon Tap* that will allow you to use it appropriately. Playback controls are on the top of your *Amazon Tap*. These playback buttons will allow you to perform functions like play or pause, forward or rewind, go to the previous song or skip forward to the next song, and adjust the volume of the speaker. All of these buttons are laid down into a perfectly flat surface up on top of the device, which provides a very slick and aesthetic feel. This side of the Tap will, of course, be accessible when your speaker is in a sling case, as there is a circular opening left on top.

At the bottom backside, you will find the power button, the 3.5mm audio input, the Micro USB port that you can use when you want to charge, and the Wi-Fi/Bluetooth button. Keep in mind that the audio and USB ports are the most important reasons that *Amazon Tap* is not waterproof-rated. It is essential that you make sure that no water gets into these ports as that could potentially be fatal for your Tap. Also, exercise caution when taking the Tap to the beach or similar outdoor locations where there is a lot of dirt, sand, or dust. You should be careful that no such particles find their way into the ports. While this may not make your Tap malfunction per se, it could interfere with connectivity later on. The buttons are fairly well isolated, though. Get to know more about them in detail below:

Power Button

The power button is used to turn your *Amazon Tap* on and off. You just need to press the power button for one to three seconds to turn it on. When the *Amazon Tap* is turned on, the power button will have a bluish tinge to it and, when it is turned off, it will be unlit. To turn the *Amazon Tap* off, press and hold the power button until you hear a tone. If you don't hear a tone, the *Amazon Tap* has gone into sleep mode.

Another thing worth noting is that these buttons will be

reachable when your Tap is in the protective sling that Amazon is selling as an addition. There is a convenient opening in the silicone case on the lower backside where the buttons are located so that you don't have to worry about practicality when taking your Tap out and keeping it protected.

Microphone Button

Once you have turned the *Amazon Tap* on, press this button whenever you want to say something. Amazon received much negative feedback from people complaining about the intrusiveness of the "always on" feature in the original Amazon Echo. With the addition of this button to *Amazon Tap*, you can now broadcast what you speak, whenever you want, by pressing the mic button located on the top corner. The reason you have to the press mic button to give a command is to save battery life.

A significant camp of consumers liked the Echo's constant awareness, though, so those buyers, in turn, complained about having to press the microphone on the Tap each time they wanted to communicate with Alexa. Although fair, this criticism is countered by the Tap's features in many ways. Pressing the button is not that big a deal when this speaker is so small, compact, light, and easily movable. Whatever the manufacturer's original intent was, one thing is for certain – this type of function greatly prolongs the battery life of the device when on the move, which is *Amazon Tap's* highest ambition.

Playback Controls

Similar to how you use the playback controls of other Bluetooth speakers and other similar devices, you can use these controls to play, pause, rewind, and fast forward your favorite tracks.

Wi-Fi/Bluetooth Button

Amazon Tap is a Wi-Fi-enabled device. You can also pair it with other devices using Bluetooth. You will find the Wi-Fi/Bluetooth button on the lower backside of *Amazon Tap*. If you need to set up the *Amazon Tap* with the application or if you want to connect to a new device, press this button for 5 seconds.

3.5mm Audio Input

Want to connect *Amazon Tap* to another compatible device? You can by using this audio input, provided you have the right audio cable. Unfortunately, the audio cable does not come with the device. However, universal audio cables are available almost everywhere, so you will have no trouble finding the right one.

It is important to know that the *Amazon Tap* does not support headphones or audio out. This is an input port only. You can, however, stream audio from other devices, such as a smartphone, tablet, or MP3 player. This is still a fairly useful feature to have because, if you have an auxiliary audio cable and a device with a lot of music in storage, it can be a very simple, old-fashioned way of playing your music on a speaker that's much better than your gadget's built-in one. This is how *Amazon Tap* fulfills the role of those older external speakers used for phones if need be.

Micro USB Port

You can attach a USB connector to *Amazon Tap* by using this USB port so that it is possible to carry your favorite music while you are on the move. This USB port will also allow you to charge your *Amazon Tap* in another way. Your other option is to use the charging cradle.

You may be wondering why you would want to charge your

Tap through the USB port when it is much easier just to put it down on the charging dock, which is always sitting at the ready. The cradle is very convenient, yes, but if you are strapped for room in your bag or whatever you may be carrying when leaving home for a period of time, the USB cable alone is easier to pack. It requires less space just to take the cable with you, as it can be coiled up and folded to take up as little room as possible.

Front Light Indicators

The light indicators will let you know what is currently going on with the speaker. After you turn it on, the blue light will come on and will pulse from left to right. If you need to speak to the *Amazon Tap*, the lights will all be lit up in blue. When they all pulse again at the same time, your request is being processed by Alexa.

If the color of your front light is orange, it means that your *Amazon Tap* is being set up with another device. It might turn blue again if it is already connected. The lights will turn red if there is an error with the request that you have made.

Charging Your *Amazon Tap*

The *Amazon Tap* can be fully charged in under four hours. You can charge your device with the micro-USB cable, power adapter, and charging cradle that came in the box with your *Amazon Tap*. It is important to ensure that you are charging your *Amazon Tap* correctly to get the most out of the lifespan of the battery.

1. Place the *Amazon Tap* on the charging cradle.

2. Connect one end of the micro-USB cable to the charging cradle and the other end to the power adapter.

3. Finally, plug the power adapter into the electrical outlet. While the *Amazon Tap* charges, you will see the power button glowing.

Another option to charge the *Amazon Tap* is to plug the micro-USB cable into the micro-USB port on the *Amazon Tap* instead of into the charging cradle and charge the device. Always use the power adapter that came with the *Amazon Tap* to charge the device. Never use other power adapters or the micro-USB ports on your computer to charge the *Amazon Tap*.

If you want to check the current battery level on your

Amazon Tap, there are three options.

1. Ask Alexa "How much battery is left?" and Alexa will then provide you with the current battery percentage.
2. Use the Settings menu in the Alexa App. This will show you the current battery percentage.
3. Use the playback controls on the *Amazon Tap* by pressing and holding both the volume up and volume down buttons at the same time; Alexa will be prompted to tell you the current battery percentage.

Andrew Mckinnon

Chapter 3
Features

Amazon Tap is Alexa-enabled, which explains why *Amazon Tap* comes with the wide variety of features that are not offered by similar Bluetooth speakers. *Amazon Tap* is designed to perform various tasks even if it has to do so simultaneously. It is much smaller in size than its predecessor, the Amazon Echo, hence it is easier to carry along with you. Your *Amazon Tap* also comes with many other built-in features, as well as the ability to connect to a multitude of other devices. The options with the *Amazon Tap* are endless, and you are sure to find something to love about this device.

The *Amazon Tap* is roughly the size of a takeout coffee cup and it can fit very well in a suitcase, backpack, luggage, or anywhere else without taking up too much space. Thanks to its Alexa capability, which we will explore in more detail later on, it can serve as a very compact portable hub through which you can run the functions of any smart home appliances you may have installed, just as with the *Amazon Echo*.

While the Tap has an edge on the Echo concerning portability, there is one feature that may make the Echo slightly more favorable for some users. This is the fact that the Echo's Alexa is listening at all times, and at quite a fair distance, so there is no need to physically interact with the device to issue voice commands. Then again, if you have privacy concerns as mentioned in the previous chapter, the Tap may be more appealing because it only listens when you press the microphone button.

Some reviewers have given *Amazon Tap* a bad rep because it has to be interacted with through the microphone button, but it ultimately comes down to personal preference. Technically, the portability of the Tap gives it an increased range as it can be brought anywhere – this is especially handy in a bigger home where Echo's Alexa may not be able to pick up on your voice everywhere in the house from its stationary position. Also, its small size hardly makes it a difficult item to carry.

Now, a smartphone can fulfill this purpose in its own way

as well, but the Tap is still a speaker superior to that of a phone when it comes to playing music. Not to mention that its nine hours of playback are sure to save you a lot of battery life on your phone.

Full Rich Sound

Amazon Tap is powered by Dolby, so it will deliver crisp sound. It gives an extended bass response that can be perfect for most songs. *Amazon Tap* also produces 360-degree omni-directional audio. Due to its cylindrical shape, sound from *Amazon Tap* can be heard all around you, giving you endless options of where and how you are able to use it.

It's worth noting that both the smaller size and the battery power capability have made it necessary to sacrifice a small degree of quality when it comes to playback, though. This mostly concerns the bass, but only as compared to the *Amazon Echo*, which, as you know, is more expensive, bigger, and powered by a cord. Of course, the sound quality is also going to be inferior to some higher-end speakers out there but, considering the *Amazon Tap's* price range, portability, and its many other capabilities, especially the capabilities of Alexa, the playback quality definitely delivers a fair punch. The *Amazon Tap* will fulfill its purpose well with any music you may use it for, while the voice sounds will also be perfectly clear and distinct.

Stream Music Directly

Want to listen to your favorite artist's music without scrolling through your library? You do not need to worry whether you have already purchased the song or not. You can ask *Amazon Tap* through Alexa to stream your favorite song through any music-streaming site of your choice. You can also control the type of music that you will hear by giving a command. We'll cover some of the mainstreaming choices and what they have to offer later on in this chapter.

Using the data contained in the individual songs, Alexa will be able to differentiate among genres, artists, albums, etc. This means that you can simply tell Alexa to play you some jazz through your *Amazon Tap*, and she will do so at random. Just like with individual tracks or artists, you can ask Alexa to stream a particular genre either from your library or from Amazon Prime Music.

Ability to Perform at Your Command

Amazon Tap will not only play music or search for your

desired queries, but it can also perform various functions. For example, you can use *Amazon Tap* to communicate with your friends, telling your location while on the move, or checking the weather before leaving home. This is all done through the Alexa app. We will cover the Alexa app in more detail later on in this book, but we can guarantee that you are going to love the options she provides through the *Amazon Tap*.

Compact

Unlike other speakers that you need to keep in one location, *Amazon Tap* is a speaker that you can bring with you anywhere you want. Want to listen to music while you bathe? As long as you have a valid and active Wi-Fi connection, *Amazon Tap* will continuously stream the music that you want to hear. Travelling? Well, you do not need to worry because you can put it with the rest of the things that you are going to pack without losing valuable space. Since most hotels offer free Wi-Fi, you will be able to use the *Amazon Tap* anywhere you go.

Of course, if you find yourself in need of Alexa's assistance while outdoors, it is also possible to tether a connection to your *Amazon Tap* by setting up a mobile hotspot on your phone. This will go through your regular mobile data plan, though, so pay attention to fees that may apply to various services that Alexa provides. This particularly concerns the services that require information to be downloaded for you through Alexa as well as streaming music from an online library.

Includes Charging Cradle

Unlike other similar devices that you can purchase online, where chargers have to be bought separately, *Amazon Tap* comes with a charging cradle. This charger allows you to listen to music or issue commands while it is charging in an upright position. If you want to charge while it is not in use, you may choose to use the USB port.

The cradle has two protruding contact points in its bed, which start conducting as soon as they come into contact with the circular conductor on the bottom side of your *Amazon Tap*. This system is designed to allow you to place the speaker into the cradle without having to worry about the direction it is facing. Whichever way the *Amazon Tap* may be rotated while in the cradle and to whichever degree, the Tap's conductor is always in contact and charging. This makes its placement very convenient and simple.

Optional *Amazon Tap* Sling

Want to personalize the *Amazon Tap*? You can do so by checking out the optional Tap sling. This sling will wrap and protect your speaker, and it comes in a choice of colors. You can choose the color that you want to personalize your *Amazon Tap* appropriately. You can select from classic colors like black and white, or you may go vibrant with blue, green, magenta, and tangerine.

The Sling includes a built-in hook that makes it easy to hang your *Amazon Tap* anywhere. This silicone casing offers effective protection while covering the speaker surface itself minimally. The bottom and top edges are protected, while the fabric cover of the speaker is held at a distance from any flat surfaces to prevent tearing or wearing of the material. If you lay your Tap down on its side on a flat surface while it is wearing its protective Sling, the fabric will never come into contact with whatever the device is laid down on. One thing to keep in mind is that the Sling will prevent the *Amazon Tap* from charging on the cradle, which means that you will either have to take it off or use your USB cable to charge the battery.

9-Hour Battery Life

Need to be somewhere for a long time when you won't be able to charge your *Amazon Tap* anywhere? Well, it will not be a problem with *Amazon Tap*, as its battery can last for up to 9 hours. Furthermore, since you can charge its battery on the go by using the cradle charger, *Amazon Tap* is certainly the right choice for consumers who have lengthy usage.

Don't forget to make use of the sleep mode to conserve battery as well. The 9-hour longevity is for continuous playback, which means that the *Amazon Tap* can take you through a much longer timeframe if you listen to music periodically and strategically. And if you combine this with a portable battery as a power source for further charging, then your wireless speaker can prove to be an invaluable addition to any sort of outdoor trip.

Tap Function

Where else can you find a Bluetooth speaker that will allow you just to touch it and use Wi-Fi without having to bring out your phone? You will not find this feature in other similar products, making *Amazon Tap* unique among its competitors. It is truly an amazing feature that you know you would not want to miss. For "tapping," you need to press the mic button found on the speaker before you can speak. Of course, unlike the Echo, this does not have an "always listening" mode. You need to tap it before it will listen and do the command that you have stated.

Pressing the microphone button is also what serves as Alexa's trigger in the case of the *Amazon Tap*, whereas the Echo's trigger is simply calling Alexa by name. This means that it's not necessary to call for her when using your Tap – simply press the button and issue your command right away.

There are many other things that you can expect to get from *Amazon Tap*. For instance, if you would like to listen to music online, you may do so through the following applications:

- Prime Music: This is Amazon's music streaming service. As long as you are already an Amazon customer and are paying for Amazon services for the year, you are going to be able to use one of its existing cloud player apps to start digging through their catalog of free music. Prime Music is full of prebuilt playlists you can choose from, and it will also recommend music for you based on your listening habits.

- Spotify: Spotify is a little different from Prime Music in that it allows you to listen to music for free. The catch is that you can only listen from your desktop, and your music will be interrupted with ads. If you choose to pay for a subscription, you will be able to sync your Spotify account to mobile devices with offline syncing privileges. Spotify also has a social aspect that allows you to see and follow what your friends are listening to and also share what you are listening to with your friends by connecting your Spotify account to Facebook and other social media platforms.

- iHeart Radio: iHeart Radio is an Internet radio platform. It will allow you to listen to thousands of the country's best live radio stations as well as its own custom artist stations. iHeart Radio is completely free but doesn't offer as much flexibility as some of the other options for streaming music. Instead of being able to choose a specific song you want to listen to, you are instead limited to what the radio stations happen to be playing.

- Pandora: Pandora is another free Internet radio option

with a little more flexibility than iHeart Radio. All you need to do is enter a favorite track, artist, comedian, or genre, and Pandora will create a personalized playlist that plays your selection, as well as similar songs.

- TuneIn: A third Internet radio option, TuneIn allows people to listen to sports, news, talk shows, and music from any location. With over 100,000 radio stations including international stations and more than 4 million on-demand programs, you are sure to find something you like with TuneIn. You can also choose to pay for a premium account, which gives you access to even more content.

If you have some songs readily available on your smartphone or gadget, you may also ask Alexa to play them for you and lo and behold: the *Amazon Tap* will play your desired music. Aside from listening to music, you will be amazed to know that *Amazon Tap* can offer you so much more, such as:

Listening to the News

Are you stuck with busy daily routine? Don't have spare time to read the newspaper in the morning? This Bluetooth speaker will give you what you need. Ask for the latest news from various reliable sites at your fingertips daily. All you have to do is listen. At the same time, you may also get weather reports. Why do you need to open a window just to see what the current weather will be when you can now ask your *Amazon Tap* about it?

Alexa can also help you keep up to date with sports, get extended weather forecasts, and find virtually any other piece of information you may desire. If you have a question about anything, just ask, and ye shall probably receive. Alexa may help with math problems, settle disputes via a coin toss, spell words, and much more. We will go into more detail in the chapter dedicated to Alexa herself.

Ask for Home Delivery

Do you want Chinese takeout? Perhaps you would like to get a box of pizza at your doorstep? It will be easy to do it with your *Amazon Tap*. It is highly likely that your other Bluetooth speakers will not be able to provide you with this type of convenience. Just imagine being able to do all of the things mentioned above without the need to use your phone or other gadgets. You can tap the *Amazon Tap*, press the button, and ask for anything.

This goes to show you that this is much more than a speaker. Rather, it is a sort of compact hub through which you can carry out many simple, daily processes and operations. Relative to particular tasks, your *Amazon Tap* can take on many roles, such as that of your phone, desktop or laptop computer, or even your own limbs if you choose to pair it with smart home appliances.

Listen to Audio Books

Amazon Tap is also a delightful help if you love to read, but there are some times when you do not want to carry a book. With *Amazon Tap*, just relax and listen to a story that you have downloaded through your *Amazon Tap*, most likely your Kindle. No need to find time to read anymore, as you can command Alexa through the *Amazon Tap* to read the book for you.

This is particularly convenient since Amazon owns Kindle and Audible, which contains an enormous library of audiobooks, easily accessible through any Alexa device you may have, including the Tap, of course. Audiobooks are especially useful if you find it easy to multitask. You can do work outside, in your garage, or in the kitchen and have your *Amazon Tap* close by your side, reading for you whatever you please.

A Note on Updates

The features that your *Amazon Tap* has are only as good as the latest update you have installed. There are constant updates being released to improve the functions of both your *Amazon Tap* and Alexa. Your *Amazon Tap* will receive these updates automatically over Wi-Fi. It is important to allow these updates to run because they improve performance and add new Alexa features. To determine the current version of the software your *Amazon Tap* is using, you can open the Alexa app and go into the settings menu in the left navigation panel. After you have selected the *Amazon Tap* from the menu, scroll down until you see "Device software version."

If you find that your Amazon Tap is not running the latest version and your *Amazon Tap* did not automatically update, you can update it yourself. First, make sure that your *Amazon Tap* is on and has an active Wi-Fi connection. Avoid saying anything to your device. Your device will automatically look for the update. Once the update is ready to install, the light indicator will turn blue, and your *Amazon Tap* will install the latest update. This can take up to 15 minutes, depending on the speed of your Wi-Fi connection. If your *Amazon Tap* is not running the update or it didn't update correctly, you can restart the device by holding the power button for five seconds, until it turns off, and then pressing the power button again to turn it back on. Once you have restarted the device, wait for the device to update again.

Keep in mind that these updates are half the magic of Alexa and similar voice assistants. You never know what the developers may come up with next and technology is spiraling almost out of control with its exponential growth. The software updates will ensure that you are keeping up with the times and always enjoying the most contemporary benefits of your personal AI. This kind of technology is still in its infancy and is being explored, so these devices may soon prove their worth even beyond what you would expect.

Andrew Mckinnon

Chapter 4
Specifications

Product Dimensions: *Amazon Tap:* 6.2" x 2.6" x 2.6"

Charging Cradle: 0.6" x 2.6" x 2.6"

Weight: *Amazon Tap:* 470 grams

Charging Cradle: 109 grams

Wi-Fi Connectivity: 802.11b, 802.11g, 802.11n standard for support with WEP, WPA, and WPA2. *Amazon Tap* can support both public and private Wi-Fi networks as long as passwords will be authenticated. Supports only 2.4 GHz band.

Warranty and Service: 1-year limited warranty and service with an option to increase limited warranty for up to 2-3 years.

Battery Life: 9 hours of continuous playing and up to three weeks on standby. Battery life may vary slightly, based on your specific device settings, usage, and environmental factors.

Audio: Dual 1.5-inch drivers and dual passive radiators for better bass sound. *Amazon Tap* also features 360-degree omni-directional audio and a seven-piece microphone array to allow you to communicate effectively with Alexa.

System Requirements: *Amazon Tap* is ready to connect to your home Wi-Fi right out of the box. The Alexa App is compatible with iOS, Fire OS, and Android. It is also accessible through your web browser. Some of Alexa's skills and services may require a subscription and be subject to fees. You will also be told before selecting a particular skill if there is going to be a charge associated with it. Standard data rates will apply when the *Amazon Tap* is tethered to a mobile device hotspot; these fees will be applied by the company from which you receive your mobile device service.

Bluetooth Connectivity: Advanced Audio Distribution Profile support is included for audio streaming from your mobile device to *Amazon Tap*, and includes Audio/Video Remote Control Profile for voice control of the connected mobile devices. Media control over Audio/Video Remote

Control Profile is not supported for users with Mac OS X devices.

Andrew Mckinnon

Chapter 5
Alexa

As mentioned previously, the *Amazon Tap* comes equipped with Alexa. Alexa is the artificial intelligence housed within the *Amazon Tap* speaker. Alexa is a sassy companion who is capable of accomplishing many great things. Getting to know Alexa and what she can do for you will be instrumental in enhancing your experience with the *Amazon Tap*.

Here we are going to cover a basic list of all the things Alexa can do for you while you are using the *Amazon Tap*. This list of skills covers just Alexa's native capabilities.

- Stream Music: You can ask Alexa to play a song, artist, album, a genre, or a playlist, and she will stream it for you.
- Read the Headlines: You can have Alexa read a news station through the music setting. She can also curate a briefing of headlines and audio clips from the news

outlets of your choice on the topics you find interesting.

- Keep Tabs on the Weather and Traffic: Alexa can let you know if there is an accident slowing down your morning commute to work. She can also give you the weather forecast. Alexa will ensure that you never leave the house unprepared.

- Timers and Alarms: You can ask Alexa to wake you up every morning. You can also have her set a kitchen timer while you are cooking or time you while you are running on your treadmill.

- Control Your Smart Home: Alexa makes controlling your smart home devices a breeze. All you have to do is pair Alexa with your smart home devices and ask her to control them as you want. The Philips Hue and Lifx smart bulbs, Belkin WeMo smart switches, and Ecobee3 and Emerson Sensi thermostats are among of the smart devices you can control with Alexa. You can use Alexa to control almost all smart home devices. If Alexa doesn't come with the capability to pair with the device in your home, you can download the skill for her; we will cover that in just a minute.

- Answer Questions: If you want to know a basic fact, Alexa can look it up for you. Or if you want help with a math problem, Alexa can solve it for you. She is also able to have conversations and tell jokes to suit all tastes. And she is well versed in movie references as well!

There are also many additional skills you can download for Alexa by using the Alexa app on your smartphone. Skills are similar to apps, and each one enables her to accomplish something new. There are hundreds of skills you can teach Alexa, and there are always new ones being developed every day. So whatever you want Alexa to do, you should be able to find a corresponding skill easily. Below is a list of some of the skill categories.

Games: If you are into gaming, Alexa can open a gateway to gamer's heaven for you! Alexa offers you access to a huge collection of games. From Abra, the character guessing game, to card games and trivia games, you are never going to find yourself bored with Alexa around.

Searches: You can ask Alexa to look up just about anything, from flight costs and movie times to the latest closing prices for stocks, gold, and oil. Alexa can also find local events, ensuring you are always up to date about what's going on around you. If you are feeling adventurous, Alexa can help you to make decisions by flipping a coin, rolling a die, or using a magic eight ball app.

Niches: You can also have Alexa give you information on games like Ark and Minecraft; she can help you tune your guitar and remember a song's name when you can only remember the lyrics, look up metric conversions, build your shopping list and even read inspirational scriptures from the Bible to you.

Some of the other things Alexa can do for you include:

- Remind you when you fed your dog
- Order pizza from Domino's
- Flag rides from Uber
- Manage your finances with a skill from Capital One
- Connect to your Fitbit so you can check things like your resting heart rate or how many steps you have taken.

Training Alexa

Although Alexa is intelligent, she may require some training before she fully understands all of your commands. Using the voice training within the Alexa app will improve the speech recognition and accuracy of your *Amazon Tap*. As time goes by and you spend more time with her, the relationship between you and Alexa will only get better.

There are a few things you need to keep in mind when you are doing a voice training session with Alexa to obtain best results.

- Speak the same way you normally would to Alexa.
- Sit or stand where you would usually talk into your *Amazon Tap*.
- You do not need to complete the entire training session in one sitting.

There are 25 different phrases you are going to say while you go through a voice training session. Your Amazon Tap is going to save every phrase you say, even if you don't finish a session. To start a voice training session, open the Alexa app. In the left navigation panel, select "settings." In the settings menu, choose "voice training," followed by "start session". Finally, speak the phrase that is displayed in the Alexa app and then select "next." If you want to repeat a phrase, select "pause" and then select "repeat phrase." When you have reached the end of your voice training session, select "complete." If, for some reason, you need to end your session early, go into the pause menu and select "end session."

Create a Recipe

One of the neat things about Alexa is that you can craft your own custom voice commands using the "if this, then that" (IFTTT) free online automation service. To do this, you have to connect Alexa to your Wi-Fi and go to the IFTTT homepage. On this page, you are going to see an option that says "My Recipes." Click on this option, and you will then see an option to "Create a Recipe." When you go into the "Create a Recipe" option, start by clicking on the big blue "this" option. Then click on Amazon Alexa and then select "Say a Specific Phrase." Now IFTTT is going to ask you for your phrase.

There are a few things you want to keep in mind when you select your phrase. First off, the phrase "Alexa trigger" is going to be the first part of the phrase when you use any of your custom commands. Alexa already assumes this, though, so you do not need to type it in your phrase. If you do type it in, you are going to have to say it twice. Your custom phrase should be written in lowercase, and you should also avoid using punctuation. Keep your phrase simple and easy for you to remember. This will save you from the frustration of saying your custom command incorrectly when you try to use it. When you have selected your phrase, click on "create a trigger."

Next, you are going to select your trigger. Navigating through the drop down menus that are provided on IFTTT and then clicking on "Create Action" can do this. After this, IFTTT will provide you with a final summary of your recipe. And if everything looks the way you want it to, you will click on "Create Recipe" and Alexa will be ready to go.

An example of a recipe you might want to set up is to have Alexa call your phone. If you have misplaced your phone, you can just say "Alexa, trigger find my phone." Alexa will then call your phone, allowing you to find it. In this case, the phrase "find my phone" is the phrase, and calling your phone is the trigger.

As you can see, you can have Alexa learn to do just about anything you want her to, to make your life easier. With the Amazon Tap, you are getting much more than just a speaker; you are gaining access to the personal assistant you didn't know you needed.

With such a level of customizability, the limits of how useful you will find your Amazon Tap are greatly extended.

Thanks to IFTTT, Alexa's practicality moves closer to being a matter of your own creativity than anything else. This is the most important way in which your Amazon Tap can be personalized, far more crucial than any protective case or aesthetic choice.

Other Things You Can Do with Your Tap's Alexa

In addition to all that we have just mentioned, there is a wide array of miscellaneous tasks that can be assigned to Alexa. Some of the things that can be done with *Amazon Tap* are simply fun, while others are incredibly useful and practical.

There is a chance you might have heard about a neat little device called Zubie. It serves as a tracking device of sorts for your car. Besides mere GPS tracking, it collects all kinds of useful data about the car's use throughout the day, like its travel route history, the exact time when the car was driven, etc. This information can be accessed in a number of ways through the Internet.

This device can now be paired with Alexa in your *Amazon Tap,* Echo or Dot. You can connect your Alexa to Zubie by installing a skill for it, just as with all the other additional skills she can learn. After you set it up, it's only a matter of asking Alexa to give you the desired information on the vehicle. You will tell her to "ask Zubie" where your car is, when it left, where it went, etc. If you happen to have more than one car in your family, individual vehicles can be given nicknames to avoid any confusion. The information you can get through this service goes to incredible depths. You can even check how much fuel is left in the car, when it sped, or when the handbrake was used. All of this is made available to you simply by asking, and it is likely that new updates will bring even more possibilities.

If you need an extra "hand" in the kitchen, Alexa can more than oblige in this area as well. You can use her to help with measurements, which she can convert easily, such as spoon measurements, cups, weight values, etc. Of course, if you need some more in-depth assistance with crafting a particular dish or meal, Alexa can read any cookbook, article, or individual recipe. You won't even have to call your mother anymore when you find yourself at a dead end in the kitchen.

Another useful thing is that you can ask Alexa to provide information on local restaurants, coffee shops, or any other business you may require. You can even use specific categories of said businesses, such as Italian food, and Alexa will search and deliver information on what you may find in your current locale. In case you need to revisit the search results, you don't have to ask again, as they will be saved and easily accessible through your Alexa app. The app's recent history will also provide other valuable information, like customer feedback and reviews of the services provided. This is a very useful system, especially if you travel and find yourself in an unexplored location.

How about getting your *Amazon Tap* to help you with your children's homework assignments? You can do this via the Teacher for Alexa Skill. This is an interactive system that assists young children with a lot of the basic math problems and tasks while keeping them entertained at the same time. This skill can be beneficial in a number of ways, as it can save you a bit of time after work and also teach your children how to interact with technology.

Another skill you may find useful is The Pianist. Alexa can use this skill to play you notes and help you tune your musical instruments or warm up before singing. You simply instruct Alexa to play you a certain note through The Pianist and off you go.

Among many other downloadable skills, you will find those that can assist with workout sessions in multiple ways, help you get ideas on what meals to prepare, and much more. There is a multitude of websites and articles that compile many of the most useful Alexa Skills, which you can sift through in your free time and grab whatever you think would suit your particular interests and daily routines. There are probably hundreds of other skills that you can get for your Tap's Alexa, and their numbers will only grow as time goes by and software becomes more sophisticated.

Andrew Mckinnon

Chapter 6
Security Concerns While Using Alexa

Alexa is capable of controlling your smart home devices as well as placing orders through Amazon for you. While this is amazingly beneficial and incredibly convenient, you need to consider the ramifications these options can have and what kind of safety precautions you should consider while you are using Alexa and your *Amazon Tap*. While there is potential for some risks, this should not deter you from using your *Amazon Tap* for these purposes. In this chapter, we are going to tell you how you can minimize security risks and never compromise your safety and privacy.

Alexa and Amazon Orders

Using the Alexa app, you can place orders using the default 1-click payment method in your Amazon account. This means that you can place orders through Amazon.com by simply using your voice and your *Amazon Tap*. You can order music with Alexa and purchase physical products. If you are an Amazon Prime member, Alexa can order select Prime-eligible items from your order history. If Alexa cannot find the product in your order history, she will offer you other alternatives. This makes it convenient to order things like laundry detergent, soap, baby supplies, and other products when you run out of them without having to use your computer. However, this also means that anyone could use your *Amazon Tap* to place a voice order. There are three things you can do to prevent this from happening.

You can find all three of these options in the Alexa App. Select "Settings" from the left navigation panel and go to voice purchasing. From here, you can change the following three settings:

1. Purchase by Voice: You can use this switch to turn voice purchasing on or off. If there are going to be other people with access to your *Amazon Tap*, it is a good idea to turn off voice purchasing. If you live alone or with family members whom you see as responsible, then this option is great to use on particular occasions only. For example, if you are hosting a party or any other kind of gathering, you can switch voice purchasing off for the

duration of the visit to make sure that nobody can play an undesirable prank or try to cause actual harm while in your home. If you happen to have jesters for friends, you can probably imagine the things they can get up to if they have access to ordering products in your name and to your address.

2. Require Confirmation Code: You can enter a four-digit code and select "save changes." Before any purchase is completed, Alexa will ask for the code. This code will not appear in your voice history. This is a good idea if you order frequently but don't want anyone who isn't authorized to place orders. This is probably the most secure way of controlling who gets to use voice purchasing. Putting a code lock on the function is ideal if you have young or teenage children in the house.

3. Manage 1-Click settings: Select "go to Amazon.com" and update your payment method and billing address.

It is also important to know that non-digital products that are purchased with your *Amazon Tap* are eligible for free returns. Simply follow the normal Amazon return procedure. While you will be required to pay the return shipping fees up front, the return shipping charges should be credited to you within seven days of receiving your product refund. This means that even if things are ordered without your authorization, while it is inconvenient, you won't lose any money in the long run.

Alexa and Smart Home Safety

The fact that Alexa can connect to your smart home devices is an amazing innovation, especially when you consider *Amazon Tap's* portability options. You can open your garage door before you get home, have your lights already on, and many other things. However, when you have connected your smart home devices to your *Amazon Tap* and Alexa, anyone who speaks to Alexa is going to be able to access those devices. This includes garage doors, lights, security systems, appliances and locks. To keep your home safe from other people who may be using your *Amazon Tap* for harmful reasons, we recommend the following safety tips:

- Follow all security instructions and recommended uses for your smart home devices. Most, if not all smart home devices come with instructions to prevent other people from using them.
- After making a request with your *Amazon Tap,* confirm whether the action was completed on the smart home

device the way you wanted it to be.
- Don't use the *Amazon Tap* for sensitive security interactions if you are in a public place or around people who you think may misuse the information. You wouldn't give your house keys to strangers and you need to take the same precautions with *Amazon Tap* and your smart home devices.
- If you notice that your *Amazon Tap* has gone missing while you were out with it, immediately go to the app and remove the smart home devices as well as any other sensitive information that someone could access from talking to Alexa.
- Take steps to ensure the safety and privacy of your *Amazon Tap* as well as safe operation of the devices you have paired with Alexa. For example, if you do not want Alexa to respond to voice commands (for example, if you have taken your *Amazon Tap* to the beach or a party) you can turn off the microphones on your *Amazon Tap*.

The chances of your security or personal information being compromised through the *Amazon Tap* and Alexa are low. It is still prudent to take some general precautions to ensure that you are protected from security threats, however low the probability may be.

As you may be thinking now, the majority of these potential problems stem from the lack of voice recognition capabilities or, rather, voice distinction and identification. This is purely a matter of technological advancement and breakthrough in the field of artificial intelligence. It is likely that our virtual assistants will only grow more clever and reliable as the years pass by and will gain the ability to recognize our voice and perceive it as the only legitimate authority. If this happens, most of these security risks will be eliminated, thus helping make our security even more automated than it is now. Either way, the steps we covered here are sure to bring the risk down to a minimum in the meantime, so exercise them and your common sense.

Chapter 7
General Precautions to Remember

Amazon Tap is built and designed to give you the best benefits possible. At the same time, it is meant to last for an extended period. Please consider some precautions that you should take to maximize the use of *Amazon Tap* and ensure that it will continuously give excellent quality.

Most reviews and feedback you can find online agree that *Amazon Tap* is well-designed and has a sturdy, quality feel to it, and you will have noticed this for yourself as well. However, it is still a somewhat delicate piece of modern technology and needs to be cared for in order for you to get the most out of it. Most importantly, it is a speaker after all, and the majority of speakers (especially smaller ones) are vulnerable to a set of universal threats, most of which can be avoided through basic common sense. Your Tap will not require you to care for it religiously, but we will look at a few basic things to note.

Your *Amazon Tap* can be vulnerable to the following things:

- Extreme Temperatures: This means both hot and cold temperatures. When you are using your *Amazon Tap* outside in the summertime, it is best to keep it out of direct sunlight. In the same vein, don't leave your *Amazon Tap* in the car when it's snowing outside. There are sensitive electronics inside the device, and those are the parts that are most vulnerable to temperature conditions. The damage will usually be permanent, especially if the Tap has been exposed to intense sunlight through much of a scorching summer day. On the other hand, a lot of smaller audio devices are affected by extremely cold conditions in a similar way, but sometimes only temporarily. You may have experienced slight distortions with things like earplugs and earphones if you have ever used them out in the cold. Although everything can return to normal after the electronics warm up again, this kind of exposure should still be avoided.
- Lack of Battery Life: You need to take care of your *Amazon Tap*'s battery. Ensure that the battery receives

a full charge when needed, but it shouldn't spend all of its time on the charging base. Again, make sure you give your Tap a proper first charge when beginning to use it for the first time. Batteries usually need to format at the beginning of their service with most devices.

- Dampness: *Amazon Tap* is not waterproof. When you are using it outdoors in the summertime, make sure that you keep it away from water. Various drinks can potentially be even worse for your Tap, particularly sugary beverages. It may be a good idea to avoid having the speaker on a table where there is dining and drinking going on. Spillages often happen and they somehow always find their way to the nearest phone, gadget, or, in this case, a wireless speaker. Put your *Amazon Tap* at a safe distance from the dining table, like on any kind of pedestal in close proximity.
- Humidity: Just as your *Amazon Tap* does not do well with water, it also does not do well in humid conditions, as this allows condensation to build up.
- Excessive Force: Your *Amazon Tap* is not intended to be hit with anything or to be thrown around. When you are carrying the *Amazon Tap* around in your bag, make sure you are conscious not to be too rough with it.

- Being Dropped from High Place: Make sure that your *Amazon Tap* is in a safe location and is not going to fall from where you set it. This is yet another reason to avoid keeping it on a table where multiple people are eating, conversing, playing cards, etc. The Tap is an upright-standing device, and it has the perfect physical disposition to get hit accidently and be sent flying off the table.
- Physical Force in General: Just as with the vast majority of speakers, bumping or rough and careless handling are big no-no's. Wallop a speaker hard enough and who knows what may come apart and get dislocated on the inside. This will usually result in permanent distortion, playback hiccups, or, in the worst case scenario, your speaker going mute. On top of taking extra care not to smack your Tap around, equipping it with a sling will greatly reduce the risk of damage.

If these things happen to your *Amazon Tap*, you may notice that there will be a change in how your speaker works. You may expect the following:

- Damage to batteries: You may find that the battery drains faster and does not have the battery life you are accustomed to.

- Not charging: If the battery is compromised, you may find that the battery no longer holds a charge after it's removed from the charging station.
- Not responding to commands: Damage to the speakers or microphones can prevent Alexa from understanding what you are asking her to do.
- Reduced speed: You may find that it takes Alexa longer to carry out the tasks you ask her to if your *Amazon Tap* has suffered any damages.
- Damaged parts: Excessive force and dropping your *Amazon Tap* may result in cracks or other damages to the *Amazon Tap*.

Amazon Tap may come with a limited warranty, but exposing the speaker to the things mentioned above will void the warranty.

There are still other tips to be remembered to ensure that *Amazon Tap* can be used for a longer period:

- Avoid placing any item on top of the device, especially something that can spill and can cause havoc to the *Amazon Tap*. If you do spill any liquid on the speaker, dry the speaker as soon as possible. Try to lessen the amount of liquid that may fall on the speaker as much as you can to minimize the potential damage to the Bluetooth speaker.

- Don't put the *Amazon Tap* where it is likely to be knocked over or otherwise damaged. Be especially cautious when you take your *Amazon Tap* outside. You don't want to get sand inside the crevices or have it accidentally stepped on. The cover of the Amazon Tap is a fabric mesh, similar to what you would normally see on a speaker, and cleaning sand and other small debris out of it will prove to be challenging.
- If you need to connect the speaker to another device, simply follow the instructions that you did the first time. Keep track of the light indicators so that you will be aware of what is happening.
- Do not try to tamper with or repair the product on your own. You might have some knowledge about repairing electronics and other gadgets, but tampering with the product if it is still on warranty will make the warranty null and void. You will also be increasing the chances of the speaker becoming more damaged in the process.
- When travelling, please ensure that the product is

stored safely in your bag. Do not keep it with other items that may spill or damage the wireless speaker in the process.

- Avoid exposing the speakers to sharp objects that may cause severe damage to your *Amazon Tap*, resulting in major malfunctioning of the device. This too can happen in transport. Again, beware of what you put next to your Tap inside your bag, suitcase, backpack, etc. Keys are a frequent culprit when it comes to causing damage to vulnerable technology and objects in general. Keep in mind the fabric cover of your speaker, which can be torn or otherwise damaged in this way.

Keeping your *Amazon Tap* Clean

To have your *Amazon Tap* last as long as possible, you should ensure that you keep it clean. The best way to keep the *Amazon Tap* clean is with a damp, lint-free cloth. Make sure that the cloth is only damp, and not wet, as you want to prevent any damages that could be caused by water. Using a lint-free cloth is also important, as you want it to pull dust out of the *Amazon Tap,* not transfer additional lint into the device.

If your *Amazon Tap* is being handled particularly often and

Andrew Mckinnon

by multiple people, then you may want to disinfect it from time to time to kill off accumulated bacteria and other microorganisms. You can do this in exactly the same way as you would wipe the speaker to clean it, except instead of damping your cloth with a bit of water, use a small amount of rubbing alcohol or similar disinfectant.

Chapter 8
Troubleshooting

Your *Amazon Tap* has been manufactured with state-of-the-art technology, and every device must meet stringent quality standards before being shipped to customers. But still, there will be times when your *Amazon Tap* encounters some problems. Please do not panic; read this part for ideas on how you can fix the issues you may encounter. You will probably find a fast and easy solution through troubleshooting. Check out the scenarios that are described below and the things that you can do to troubleshoot your device efficiently.

Amazon Tap Does Not Pair with Your Device

In case of trying to pair your speaker with your device, here are some of the things that you can check:

- Try to setup *Amazon Tap* with your device manually by pushing the Bluetooth/Wi-Fi button for up to 5 seconds. Using your device, go to "Settings" and tap on "Set up the new device."

- Check if the OS version that you have on your smartphone or gadget is compatible with Alexa. If not, you may need to update.
- Ensure that your *Amazon Tap* receives regular software updates. Software updates are meant to improve functionality and fix issues. If your *Amazon Tap* doesn't get the latest software updates, it may malfunction.
- Check if your device is compatible to be paired with Alexa Voice Service. Remember that early smartphones as well as some Kindle versions are not compatible with Alexa and cannot be used to pair with *Amazon Tap*.
- Restore factory settings, as you might have changed the settings of the speaker unintentionally, which might cause the disconnection with your device.
- Please make sure that you are connected to the Internet. It will not pair your device with Alexa Voice Service without an Internet connection.
- Restart your device and restart *Amazon Tap*.

Amazon Tap Does Not Respond to Your Commands

- Check if the speaker is turned on.
- Make sure that you tap the button located on the front portion of the *Amazon Tap* before you state your

command.

- As we have said before, make sure that you are running the latest version of the software on your *Amazon Tap*.
- Check the indicator lights to see if they show that *Amazon Tap* is responding to your request.
- Revise the way that you speak your command. Make sure that you use simpler and shorter terms to make Alexa work.
- Restart the speaker. There are times when your speaker simply wants a simple restart, and it will function normally again after you have turned it on and off.

Alexa Doesn't Understand You

- Make sure that your *Amazon Tap* is at least eight inches away from walls and other objects that might cause interference, such as baby monitors, and microwaves. If your *Amazon Tap* is on the floor, try moving it up higher.
- Make sure that there is no background noise, and speak naturally and clearly to Alexa.
- Repeat your request or question. You can also rephrase it to make it less general. Check the Alexa app to see what Alexa heard; this can make it easier for you to troubleshoot the issue. To see what Alexa heard, go to

the Alexa app and, from the home screen, select "More" at the bottom of the interaction card. This way you can read what Alexa heard, listen to the request and provide feedback.

- If the issue seems to stem from Alexa's inability to understand you, use voice training to help Alexa learn your specific tones and voice inflections.

Amazon Tap Does Not Turn On

- Press the power button and wait for five seconds for the lights to be turned on. If the light is blue, you have successfully turned it on; if it is red, it means that you do not have enough charge in the battery for *Amazon Tap* to operate.
- Charge the *Amazon Tap* using the cradle charger and see if it will turn on while it is charging.
- Charge the *Amazon Tap* using the USB port.

Doing a Factory Reset on Your *Amazon Tap*

If restarting your *Amazon Tap* doesn't resolve any of your issues, you can do a factory reset of your device to see if that fixes the issues.

- Press and hold the Wi-Fi/Bluetooth button as well as

the Previous button simultaneously for 12 seconds. The light indicators will turn orange and then blue.
- Wait for the light indicators to turn off and then turn back on. When the light indicators turn orange, your *Amazon Tap* has entered Setup mode.
- Open the Alexa App and connect your *Amazon Tap* to a Wi-Fi network. Then register it to your account and follow the steps you did when you originally set up the device.
- In the case of any other problem that is not mentioned above, contact Amazon technical support team by visiting the Amazon website.

Andrew Mckinnon

Conclusion

Thank you again for downloading this book!

Amazon Tap is a great addition to Amazon family of technological advancements. It is designed to have all of the features of Amazon Echo and Amazon Dot, along with a few additional features that make it a handy device to carry when you're on the go.

As you have seen, while *Amazon Tap* takes a somewhat different approach to providing Alexa's services via its button tap function, it still offers most of the same functionality, with the addition of being portable and completely wireless. While choosing between the Tap and the Echo essentially comes down to personal preference, the Tap is still a cheaper option with a few key features unique to it. It will fulfill the role of a speaker either through Bluetooth pairing with your device or by way of connecting a device to it through an audio input cable. And beyond that, of course, its playback capabilities are the least of its advantageous features.

If you were an Echo user prior to procuring your *Amazon Tap*, there are a few things to get used to, as you have seen. But, if you look at both devices individually and judge them on their own merits, it's apparent that each has an edge on the other in some respects. However, the price difference and the portability of the Tap may be just the things to sway a lot of people towards *Amazon Tap*. It all depends on what you are looking for in a Bluetooth speaker, and this book has hopefully provided the information necessary for you to make that call.

I hope all the things in this guide help you use your *Amazon Tap* effortlessly from your first try! We've covered the different angles to successfully using your *Amazon Tap* and getting the most out of its features while ensuring it lasts to its full potential.

This book has shown you some examples of all the things you can do with the *Amazon Tap*. We covered how to take care of your *Amazon Tap* as well as some steps you can try if you ever experience issues with your *Amazon Tap*. Not only have you learned about the *Amazon Tap,* but we have also given you some useful information about Alexa. Now you know what Alexa is capable of as well as how to train her to achieve the best results. By following all of the tips and tricks that are outlined in this book, you will certainly have a great experience with Alexa and your *Amazon Tap*. All you have to do now is get to know Alexa better. Experiment with Alexa to figure out how she can improve your life. Before you know it, Alexa will become an indispensable part of your life. You are now ready to head out and enjoy your music anywhere and everywhere you want to go!

Finally, if you enjoyed this book, then I'd like to ask you for a favor, would you be kind enough to leave a review for this book on Amazon? It'd be greatly appreciated!

Andrew Mckinnon

www.ingramcontent.com/pod-product-compliance
Lightning Source LLC
Chambersburg PA
CBHW060405190526
45169CB00002B/754